DATE DUE

APR 1 2			
MAR 2 2 1989			
MAY 2 3 1989			
MAY 1 1990			
MAR 1 3 1991			
JUL 1 8 1991			
DEC 1 0 1991			
APR 1 9 1995			
DEC 1 7 1998			
DEC 1 4 2001			
MAR 2 3 2005			
NOV 1 9 2014			

HIGHSMITH 45-220

What is ecology?

Every living thing that is known to exist is found on one planet, the Earth. They all share this planet, from bacteria too small to be seen without a microscope to the giant redwood trees and the whales of the oceans.

All the living and non-living things that surround such a plant or animal are called its environment. For example, the environment of a plant includes the soil, the water and foodstuffs in the soil and the air the plant is growing in. Rainfall and temperature may affect the life of the plant as well as other plants that may compete for water and food. There may also be animals that eat the plant and some that may help it to reproduce. All these things make up the plant's environment. The science that looks at the ways in which plants and animals affect their environment, and are affected by it, is called "ecology."

AIR ECOLOGY

Jennifer Cochrane

Series Consultant: John Williams, C.Biol.,M.I.Biol.
Series Illustrator: Cecilia Fitzsimons,B.Sc.,Ph.D.

The Bookwright Press
New York · 1987

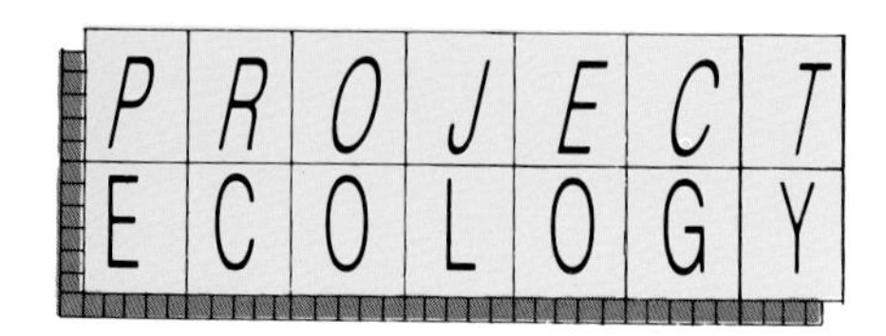

Air Ecology
Animal Ecology
Land Ecology
Plant Ecology
Urban Ecology
Water Ecology

First published in the
United States in 1987 by
The Bookwright Press
387 Park Avenue South
New York, NY 10016

First published in 1987 by
Wayland (Publishers) Ltd
61 Western Road, Hove
East Sussex BN3 1JD, England

ISBN 0-531-18151-0

Library of Congress Catalog
Card Number: 86-73066

Printed in Italy by
G. Canale & C.S.p.A., Turin

Jennifer Cochrane is the author of many titles on ecology and the natural world.

John Williams is a former teacher and a school science advisor.

Cecilia Fitzsimons is a science-trained artist who specializes in natural history and biological illustrations. She has a doctorate in marine cell biology and has taught science.

Cover: bottom *seeds blowing from a dandelion*, left *a fruit bat in flight*, right *air pollution from factory chimneys.*

Frontispiece: *air pollution from industry in New South Wales, Australia.*

Contents

1. A blanket of air

In these days of space exploration, everyone knows that the planet on which we live, the Earth, is one of at least nine circling the Sun. Each one is different. Mercury, nearest to the Sun, is a small, rocky planet without a surrounding blanket of gases. Such a blanket is called an atmosphere. Venus, Earth's nearest neighbor, has a very hot, choking atmosphere made up of the gas carbon dioxide. Mars, our other neighbor, has a thin atmosphere of this gas, but it is very cold since it is farther from the Sun than the Earth.

The air around us is made up of a mixture of gases. Nearly 80 percent of it is the odorless, invisible nitrogen and about 21 percent is made up of oxygen. The remaining part of the air, less than 1 percent, consists of carbon dioxide and gases like argon and neon. In addition to these the atmosphere contains water and a great deal of dust and other solid things floating in the air. The Earth's is the only such atmosphere in the solar system. It supports all life on the planet.

The Earth's satellite, the Moon, has no atmosphere and we can see what the Earth would be like if it had no wrapping of gases around it. Where the Sun shines on the Moon's surface it is hot enough for water to simmer. In shadow, it is far colder than

The cratered surface of the Moon with the Earth in the background, as seen from a manned space flight.

Antarctica on Earth. Also, there is nothing to stop meteorites and the billions of tiny dust particles in space from hitting the surface.

The Earth's atmosphere absorbs much of the Sun's heat and keeps the planet warm where the Sun is shining and at night stops the warmth from escaping back into space. The space dust and most of the meteorites that enter the atmosphere burn up before they can reach the surface. About a million meteorites enter the atmosphere every day and are sometimes seen as "shooting stars" in the night sky as they are burned to dust by friction with the gases.

The Earth's atmosphere is not just a single layer of gases. We can imagine that it is made up of a number of layers. About half of all the air is in the troposphere, the layer closest to the planet's surface. All of the weather is in the troposphere and it is the layer where life is found.

Much higher up is the stratosphere and here is found a layer of the gas ozone. A great deal of the ultraviolet radiation from the Sun is absorbed by the ozone, and without this protective skin much life on the planet would not be possible.

In the upper part of the atmosphere is the ionosphere. The air in this region is very thin indeed, but it is an important layer for humans. The ionosphere reflects radio waves back to the planet's surface, around the curve of the Earth.

The Earth's atmosphere makes life possible on the planet. It keeps the temperature even, keeps out solids from space and harmful rays from the Sun, and supplies the gases that living things need to breathe. No other planet has an atmosphere like ours and, as far as we know, all other planets are lifeless.

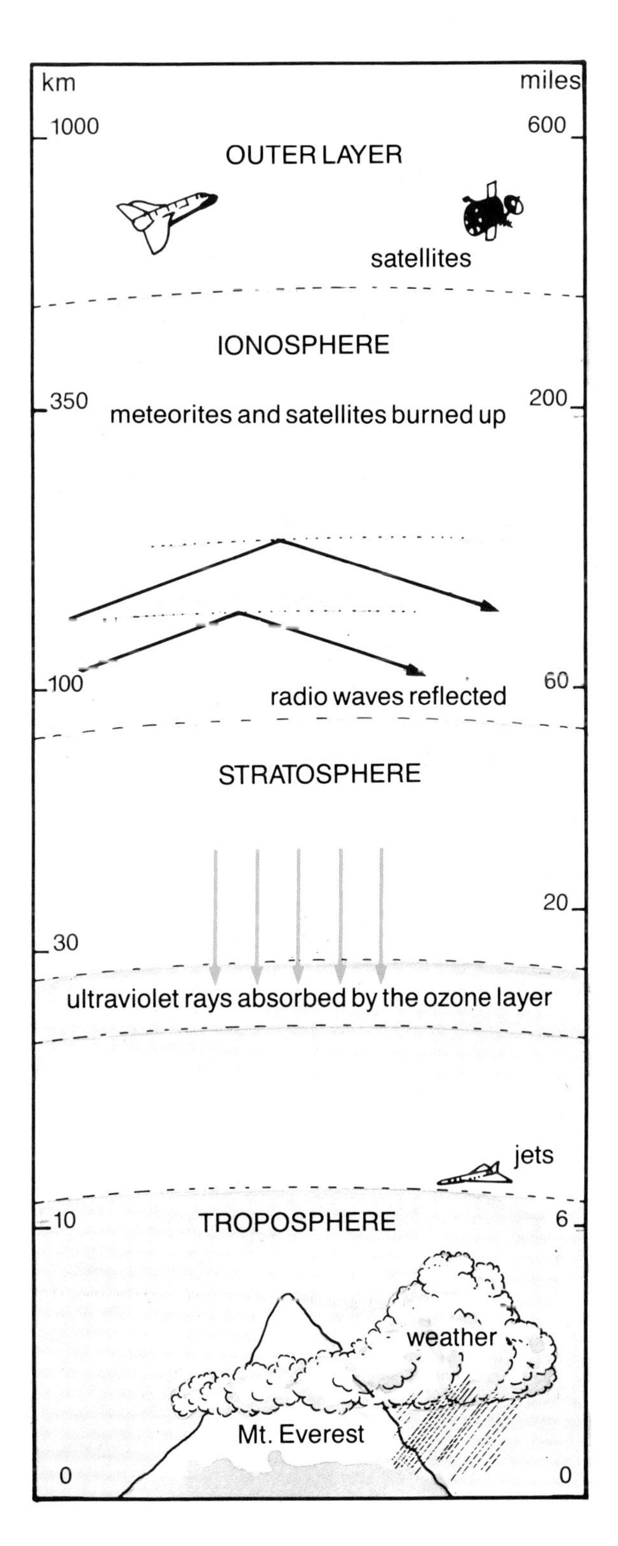

A cross section of the Earth's atmosphere.

2. A cycle of gases

The planet's atmosphere has not always been the same. Over 3 billion years ago, when the Earth was still cooling after its birth, the gases swirling about the hot surface were probably carbon dioxide, hydrogen, methane and ammonia, among others. There would have been little oxygen.

The atmosphere began to change when the first green plant life appeared about 2,700 million years ago. These were simple plants like algae that lived in water and were able to make their own food. This is because green plants contain a chemical called chlorophyll, which enables them to use the Sun's energy to convert water and carbon dioxide into food. This is called photosynthesis and is very important since oxygen is released as a result.

Over millions of years the amount of oxygen in the atmosphere steadily built up because of the green plants. It was only after there was sufficient oxygen in the air that animals could develop on the planet.

Oxygen is essential for all animal life, including ours. In this process called respiration (breathing) oxygen is taken into the body and carbon dioxide is released in return. Green plants "breathe" when there is no light. Eventually the production of the two gases on Earth reached a balance and the mixture that we depend on today has not changed for a long time.

Carbon dioxide is also released by the decay of dead plants and animals and by the burning of fossil fuels like coal and oil. The oceans absorb vast amounts of this gas and also release it so that carbon dioxide is passed between the air, plants, animals and the oceans. With oxygen passing from plants to animals, a cycle of gases between living things and the environment is created.

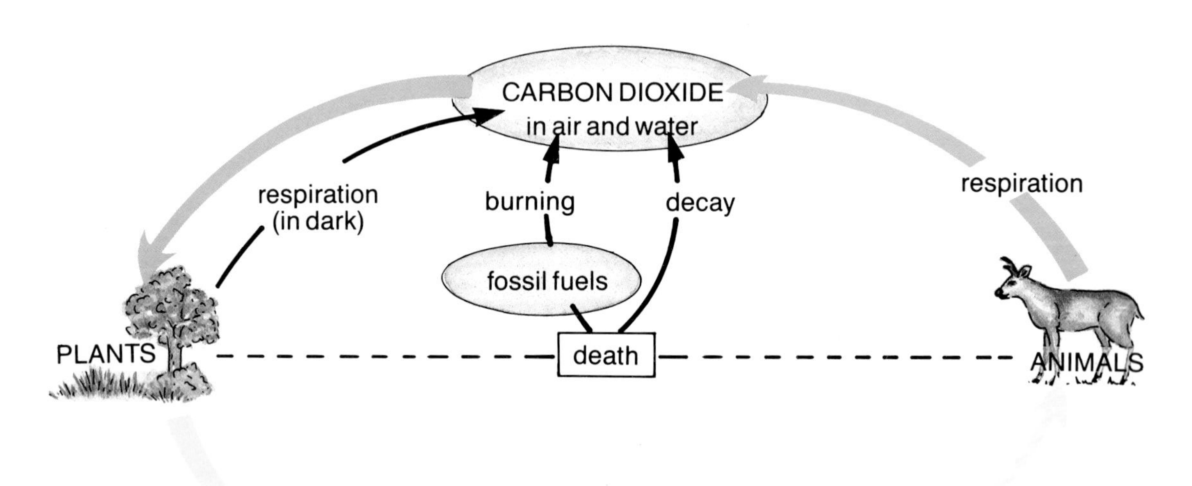

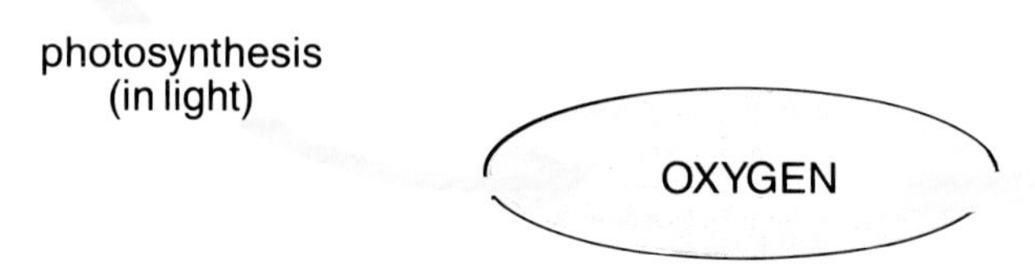

The gases that are essential for life on the planet are passed between living things and the environment.

Activity: A gas released by plants

What you will need

Some dilute bicarbonate indicator solution, a little pond weed, a tall beaker or jar, a small funnel and a test tube. You will also need three small pieces of plasticine, a straw and a wooden splint.

Arrange the pieces of plasticine so that the base of the funnel will sit on them. Put the weed between the plasticine pieces and fit the funnel over it. Carefully fill the beaker with the bicarbonate solution until the funnel is well covered.

Fill the test tube with some bicarbonate solution and put your thumb over the end. Carefully turn the test tube upside down and with your thumb still in place put its mouth under the solution in the beaker. Make sure there is no air in the test tube and remove your thumb, putting the test tube over the funnel as shown.

Next, gently blow into the solution using a straw. The bicarbonate solution turns yellow in carbon dioxide. The solution is now charged with the carbon dioxide from your breath.

Leave the experiment in sunlight for a few hours, frequently checking it. What happens to the color of the solution? Is it necessary to "recharge" the solution with carbon dioxide?

Bubbles of gas should appear on the weed and collect in the test tube. When it is full of gas, light a wooden splint. Blow out the flame and while it is still glowing remove the test tube and hold it upside down as you put the splint into it. Does the flame reappear? If it does, this is a sure sign that oxygen is there.

What this shows

The weed has taken in carbon dioxide from the solution and, through photosynthesis, has given out oxygen.

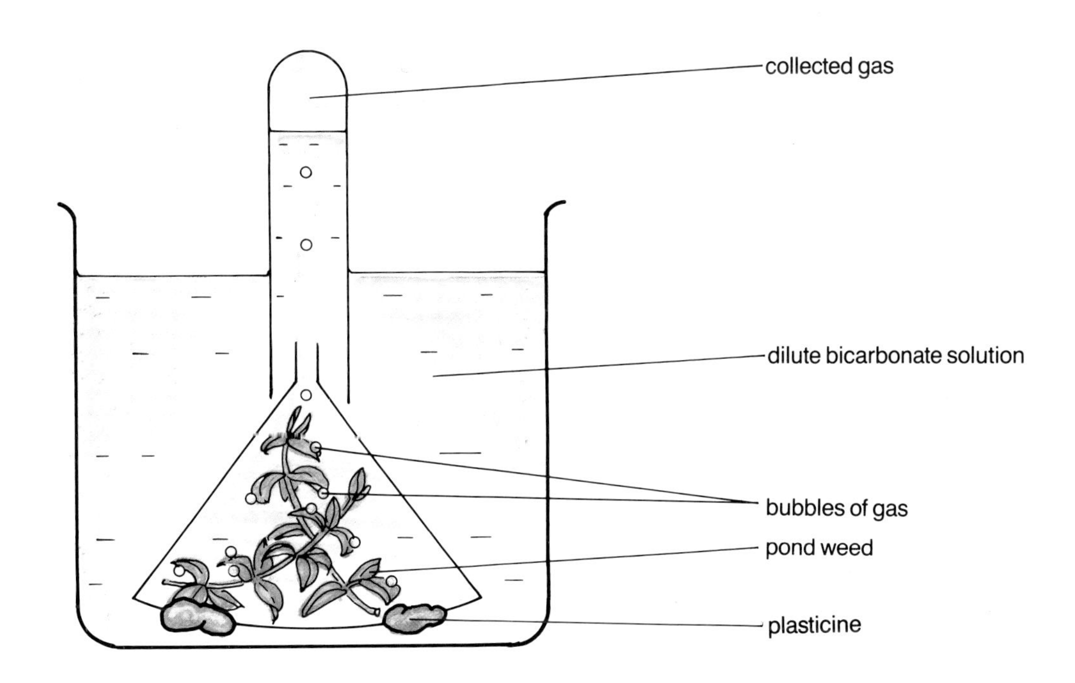

3. The greenhouse planet

Most of the Earth's heat comes from the Sun. The Sun produces heat, light and other forms of energy that travel through space as rays, or radiation. Only a tiny fraction of the Sun's energy reaches the Earth and even less gets to the planet's surface.

The energy may be reflected back into space or it may be absorbed by the planet. For example, a lot of the energy is absorbed by the atmosphere and much is scattered by dust particles. In addition, about one-third of the energy is reflected by clouds and the surface of the Earth.

Fresh snow is such a good reflector that most of the energy falling onto it is sent back into space. On the other hand, the tropical rain forests, with their very dense amounts of plant life, reflect only one-tenth of the energy they receive, the rest being absorbed. The Earth must also give out (radiate) the same amount of energy that it has absorbed, otherwise the planet would have burned up long ago.

The carbon dioxide in the atmosphere absorbs energy and affects the temperature of the planet by what is called the "greenhouse effect." A greenhouse works by letting the Sun's heat rays pass through the glass to heat the soil and plants. These in turn give out heat rays, but since the soil and plants are at a much lower temperature than the Sun, the heat they radiate is rather different (infrared radiation) and cannot pass out through the glass. The air inside heats up as a result.

Fresh snow, like this at the North Pole, reflects about 85 percent of the sunlight falling on it.

The carbon dioxide in the atmosphere acts rather like the glass in a greenhouse, letting the heat from the Sun through to the Earth but preventing some of the radiated heat from escaping. This keeps the Earth at a higher temperature than it would otherwise be.

In the "greenhouse effect" the Sun's rays pass through the atmosphere but the radiated heat from the Earth is trapped by the carbon dioxide, causing the Earth to become warmer.

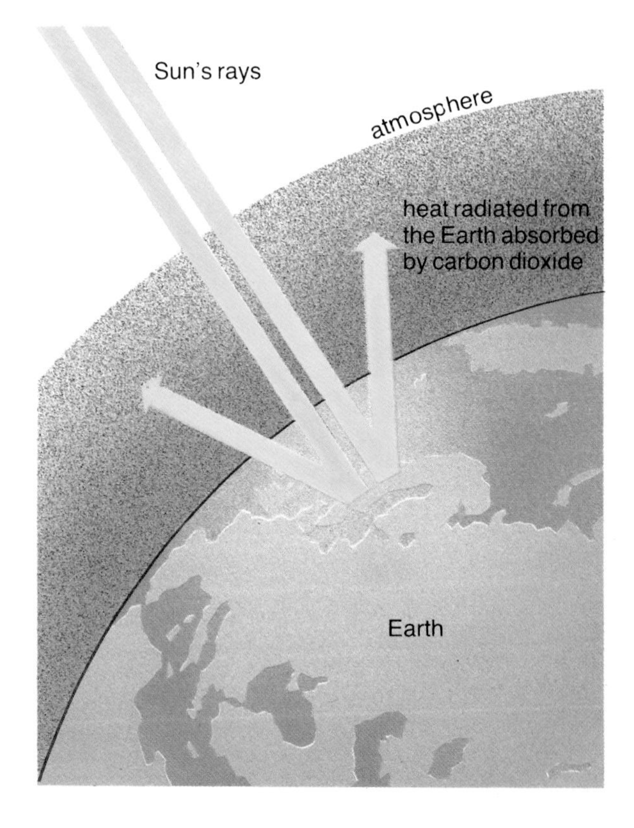

Activity: Absorption and reflection

What you will need

An electric lamp, black and white poster paint and brushes.

Hold the back of your hand close to the switched-on light bulb. You will feel the heat radiation from the bulb. Be very careful not to burn yourself. You will soon have to move your hand.

On the back of one hand paint a large black patch and on the other a white patch. Hold the backs of your hands close to the bulb at the same distance from it. Does one hand feel hot before the other? Does the hand painted black get hot quicker than when there was no paint? Are you able to hold the hand painted white near to the bulb for longer? Which color absorbs heat the best and which reflects it the best?

Why do you think people in hot countries often wear light-colored clothing?

What this shows

Light-colored objects absorb little heat and reflect a lot. Dark colors absorb more heat and reflect less.

The great forests of the world, like this rain forest in Borneo, absorb about 90 percent of the Sun's energy falling on them.

4. Light into food

The energy stored in your body comes from the food you have eaten. Humans, like all animals, cannot make food in their own bodies and so have to eat either green plants or animals that have eaten plants. The energy taken from the plants can be traced back further: it was the Sun's energy that enabled them to make their own food during photosynthesis. So the energy we use for breathing, growing and moving has come originally from the Sun.

The atmosphere must allow enough light to reach the surface of the planet for plants to produce food and oxygen through photosynthesis. There must also be enough green plants to trap this energy so it can be transferred to animals which depend on them.

Where there is much plant life, such as the regions around the equator, animal life is found in great numbers. Where there are few plants, such as deserts and ice-covered areas, the number of animals is smaller as a result. So green plants are essential for all other forms of life on the Earth.

Very little of the energy reaching the Earth in the form of sunlight is used by green plants in photosynthesis. Of the energy that is absorbed by them, only 1 or 2 percent is actually stored in the products of photosynthesis, such as the food starch. So plants are not very efficient at transforming light energy into food energy. This does not matter, though, since there are so many green plants. By weight, they are 99.9 percent of all living things, with animals and colorless plants making up the rest.

It is still important for humans to take care of this plant life though. Humankind has put smoke and dust particles into the air and to continue to do so may cut down the light reaching the surface of the Earth and the plants. Humans can now change the climate of the planet by accident and this could also affect plant life, the life we all depend on.

Green plants, like grass, use the Sun's energy to make food. This energy can then be passed to all animals, including humans.

Activity: Plants make starch

What you will need

Two potted plants with large leaves, stiff cardboard to make stencils, some ethanol (methylated spirits), iodine solution, a water bath with a lid, a beaker, a tripod, a bunsen burner, a pipette, some paper clips and a long pair of tweezers.

Put the two plants into a dark cupboard for at least twenty-four hours. This will make sure that there is no starch left in the leaves. Then take out one of the plants. Take a piece of cardboard and cut out a simple shape. Fix your stencil to a leaf with some paper clips. Repeat this for a number of leaves and keep the plant in sunlight for a day.

After a day remove the leaves with the cardboard and take off the stencils. Now test these leaves to see if they have made starch. First, the chlorophyll must be removed. Put the leaves into a beaker of ethanol and place this beaker in the water bath as shown. With a low flame from the bunsen burner, heat the water and the ethanol will soon boil. Why is it important not to heat the ethanol directly?

When the chlorophyll has been removed turn off the heat, remove the lid and carefully take out the leaves with the tweezers. Lay a leaf on some paper and drop a little iodine solution onto it with the pipette. If starch is present the leaf will turn blue-black. Is there a shape of color on the leaf? What does this mean? Repeat the test for a leaf that has been in the dark.

What this shows

Leaves need light to make their food.

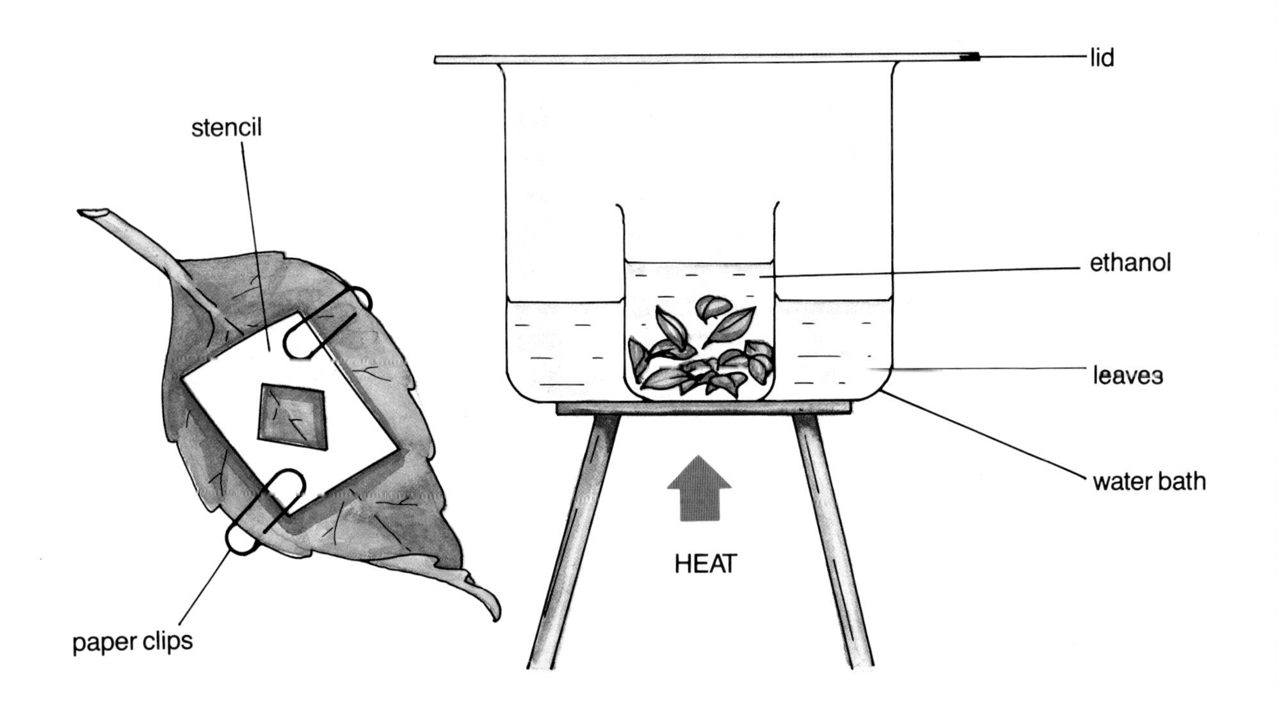

5. The air system

The air does not remain still but moves around the planet as the winds. Many of the Earth's winds depend upon the rule that warm air rises. This is because warm air expands and becomes less dense than the surrounding air. As a result it rises, rather like a cork released from the bottom of a tank of water.

Sea breezes show how this works. In the Sun both land and sea warm up but land heats up quicker than water. The warm air above the land rises leaving an area of "low pressure" over the surface where there is less air. Cool air from the sea then moves in to fill this area and the result is an "onshore" breeze. Such movements of air are called convection currents.

Much larger-scale winds that may stretch for hundreds of miles are made in a similar way. The equator receives more of the Sun's energy than the poles since the Sun is directly overhead. The air at the equator heats up, expands and rises leaving a low-pressure area. At the cold poles the air cools and shrinks, falling to the surface to give an area of high-air pressure. Warm air from the equator then moves to fill the space left by the fallen air above the poles. This extra air cools and sinks, pushing away the cold polar air, which is drawn to the equator to fill the space left by the heated air.

These winds do not move from the equator to the poles in straight lines since the planet is spinning, dragging the air with it in great spirals.

Humans must be very careful about what is introduced into the air system. Water and heat are important things carried by the winds around the world, but pollution can also be taken with them.

In the heat of the Sun, the air over the land warms and rises. The space left behind is filled by a breeze from the sea and this is called a convection current.

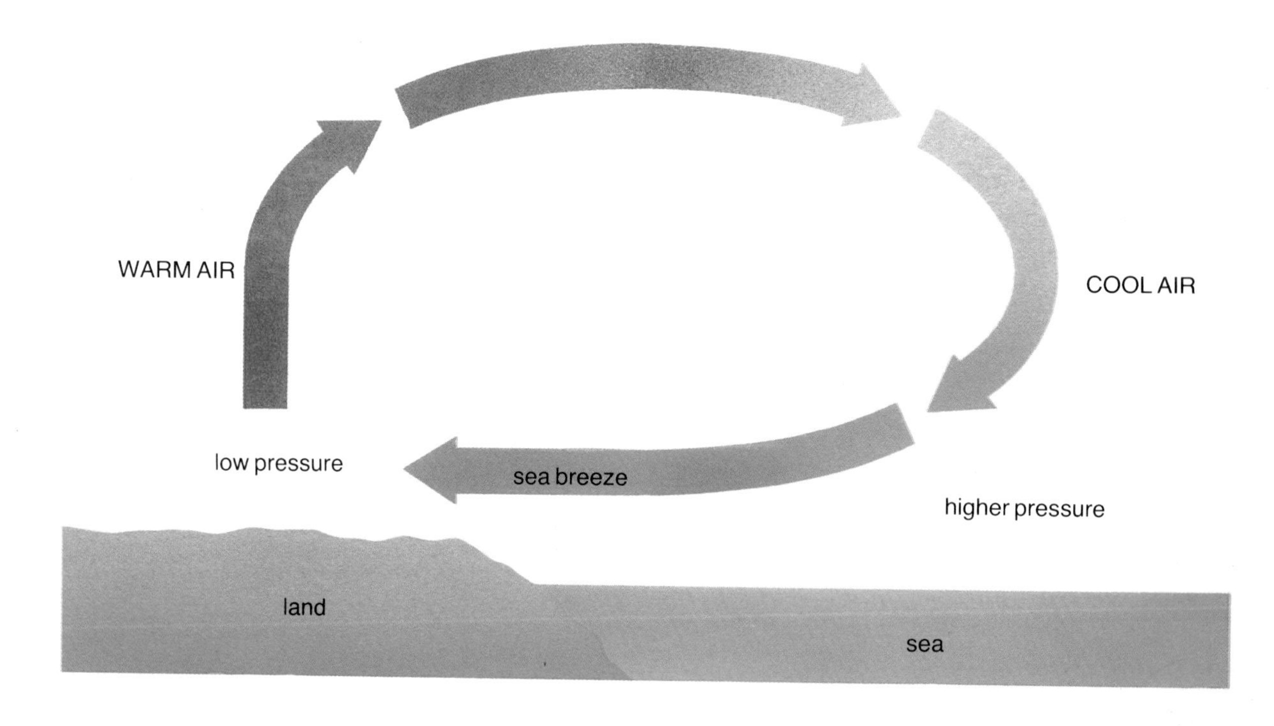

Activity: Make a hot-air balloon

What you will need

A thin, lightweight plastic bag such as those used for wastebasket liners and a hair dryer.

Hold the bag with its neck wide open and use the hair dryer to blow hot air into it. The bag will slowly fill with air, or inflate. Hold the bag until you can feel it trying to lift, then remove the hair dryer and let the bag go.

Why does the bag rise? How high does it go? Why is it important to use a very lightweight bag?

Find other lightweight bags of different sizes and repeat the experiment. Which travels higher, a large or small bag?

What this shows

When air is warmed it rises, taking with it lightweight substances. This is how a hot-air balloon rises, since it traps a large amount of heated air which produces a lift.

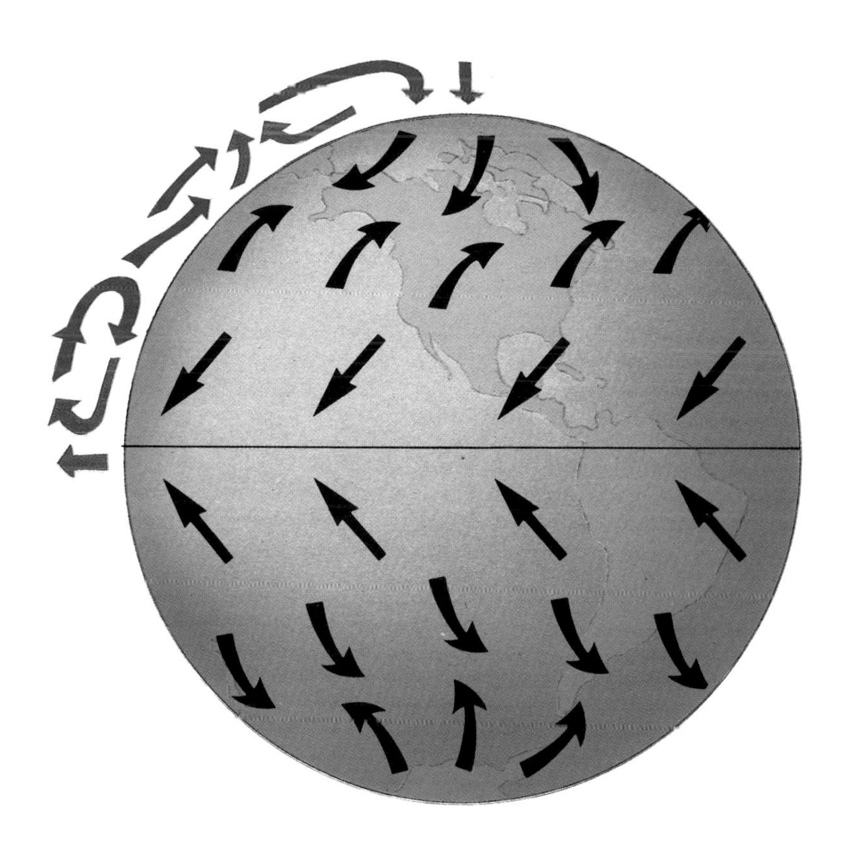

The world's constant winds are caused by convection currents. These currents are shown in the top left-hand corner where warmed air rises, traveling from the equator toward the poles, while cooled air sinks and travels toward the equator.

6. Moving the water

Any substance can be either a gas, a liquid or a solid. At normal temperatures water is a liquid, but above 100° C (212°F) it all becomes water vapor. Yet even at lower temperatures some of the water passes into the air as vapor. This is called evaporation. Liquid water needs heat energy to turn it into vapor and the Sun provides most of this.

All land-living plants and animals depend on fresh water, which falls as rain or snow. Some of it drains into lakes and is absorbed by the soil. However, the water will reenter the atmosphere by evaporation from the plants, the land and the areas of water. The vapor will eventually turn back into a liquid and fall, so continuing what is called the "water cycle."

Warm air can carry more water vapor than cold air. So, in cold temperatures, the vapor turns to liquid, which may stay in the air as tiny droplets, so light that they continue to float as clouds. The droplets in the clouds may join together so that they are too heavy to float and fall as rain.

The movement of water is very important. For example, the summer monsoon in India brings vital rain for the growing of crops. The central plains of India are heated during the summer and the air rises as a result. Air that is charged with water is drawn in from the Indian Ocean to fill the gap left by the risen air and torrential rains result.

Water vapor is also carried from the equator to the colder regions where it turns to liquid and forms clouds. When vapor turns to liquid water, there is a release of energy as heat. This is why the poles are warmer than they would be if there were no movement of water

A fog of tiny water droplets over a rain forest in Surinam, South America. Some of the water will evaporate from the ground and the plants back into the atmosphere. Here it will turn back into a liquid and fall again.

Activity: Water lost and gained

What you will need

Water and several containers of different sizes such as large and small plates, cans and trays. A beaker or jar and some ice will be needed for a second experiment.

Fill the containers with water and put them in the Sun or in a warm place. Mark the levels of water on the containers and leave them for a few hours. Check the levels after each hour.

Have the water levels changed? Where has the water gone? Which types of container lose the most water? You could also measure the size of puddles of water after it has rained and check their size during the day.

Put some ice into a beaker and fill it with cold water. Place the beaker in a tray and watch what happens. What appears around the sides of the beaker? Where has it come from? Why is it that your breath turns to mist on a cold day?

What this shows

Water does not have to boil for it to turn into vapor, or evaporate. Those containers that had the largest open area lost the most water. This shows that water evaporates at the surface of the liquid.

If air is cooled it will give up some of its water as droplets. In this case water vapor turns to liquid and this is called condensation.

A lagoon in Australia that has almost dried up during a period of drought.

7. Air and climate

Temperature, sunlight and the winds carrying water vapor and rain are responsible for the climate of a particular area. For example, the regions around the equator are the hottest parts of the world with very heavy rainfall. Since warm air can hold a lot of water vapor the atmosphere is very humid, or damp. These conditions support the rain forests. The polar regions receive less heat and are therefore cold, with little or no vegetation.

The land may also have an effect on the climate. Mountain ranges force the winds to rise into cooler regions so the water vapor turns to rain or snow. This results in the air being cool and dry when it crosses the mountain and moves down the other side. As it falls it warms and absorbs water vapor. This causes a dry area called a rain shadow to be created in which rain rarely falls. The Gobi desert in Asia and the Great Basin in the western U.S. are rain-shadow areas.

Yet living things also play a part in forming the climate in which they live. For instance, plants absorb some of the Sun's heat and give out water vapor, and humans have affected the climate of cities with smog. This is a harmful mixture of smoke, gases and tiny water droplets which have been trapped near the ground.

Normally, air that is warm at the Earth's surface is constantly rising and cooling, to be replaced by air from elsewhere. In these conditions smoke is quickly removed. Sometimes, though, these air currents stop flowing because a layer of warm air develops above the cooler air at the surface and below the upper cold layer. In this case, the lower layer of air is fairly still and smog forms, trapped near the ground, especially if the city is in a valley or in a depression in the land.

Polluted air trapped over a town.

Activity: Making a climate

What you will need

A fish bowl or large jar, some kitchen foil, ice cubes, some paper tapers and matches.

Put a little water in the bowl and then pour it out so that a few drops of water remain on the insides. This is to make the air in the bowl moist. Drop a lighted taper into the bowl. Quickly cover the top with aluminum foil so that it fits securely and add the ice cubes. Then watch what happens inside the bowl.

What happens to the air as its temperature falls? What has caused this, remembering that water droplets were left inside the bowl?

The smoke rises since the air is warmed but it meets the layer of cold air under the ice. Smog which cannot escape forms here and as it cools it falls back to the bottom of the bowl. Why does smog like this occur more in cities that have rivers running through them? Why does smog happen more in the winter?

What this shows

When smoke particles and the moisture in the air are trapped near to the surface they combine to form smog. This happens especially when the air contains a lot of water and when the temperature is cold enough to turn the vapor into a mist.

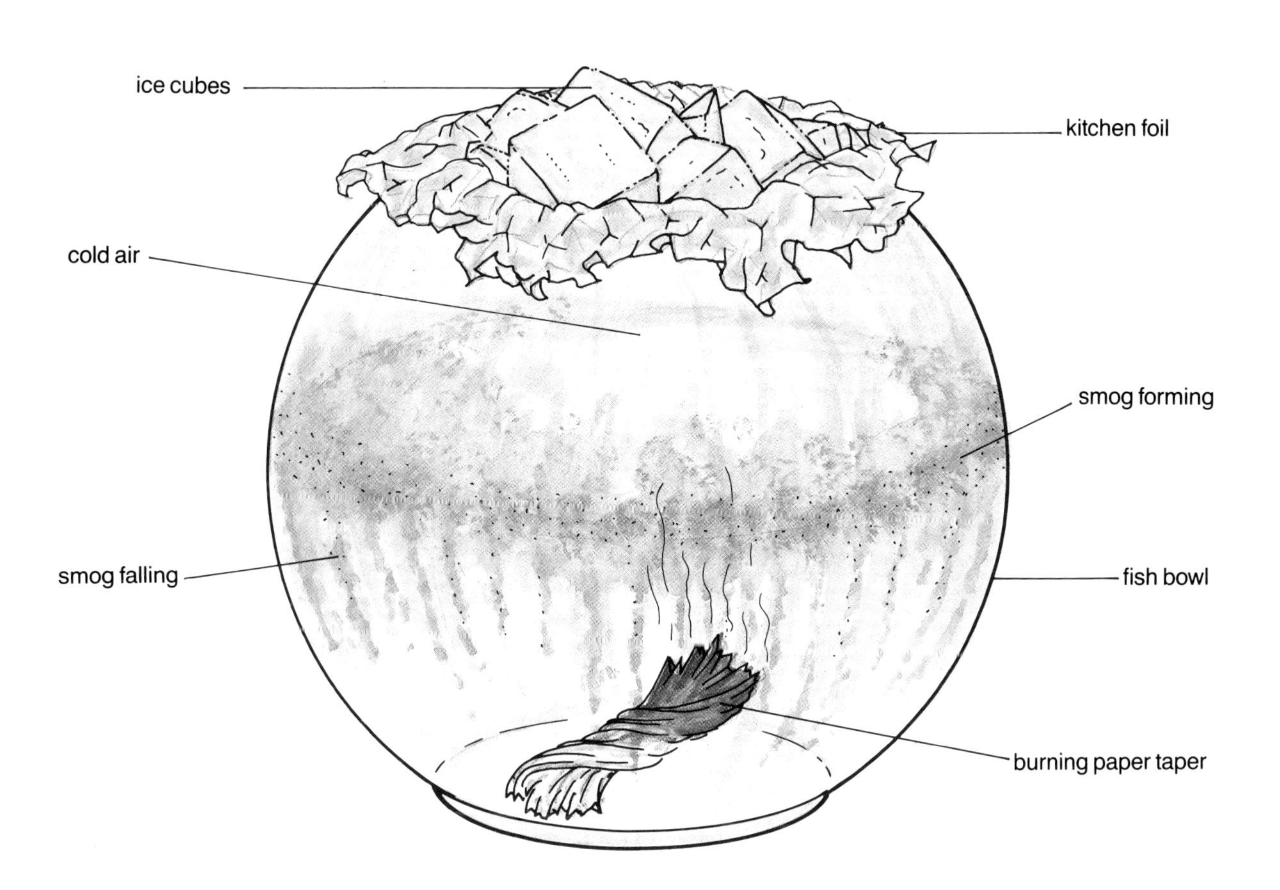

8. Wind as an enemy

The wind can be both a friend and an enemy to living things. For example, fierce winds can damage vegetation and blow away the soil that plants need to obtain their water and nutrients.

Many strong winds begin as air currents inside clouds. Sometimes conditions make the air moving up and down in the clouds begin to spin. This gets faster until a funnel of air dips down to touch the land. This is a tornado, and immense damage is caused by whirling winds sucking debris into the funnel and carrying it into the air.

Hurricanes are giant versions of tornadoes, sometimes 1,000 km (620 mi) across. They are "whirlpools" in the atmosphere and cause great damage. However, they move quite slowly and it is fairly easy to tell where they will go so that warning can be given.

Plants suffer the most damage from drying out when winds increase the evaporation from the leaves. In this way, leaves that face into the wind will be killed and branches on the other side will be shaped like flags in the wind. Some plants have adapted to this by developing strong branches or by hugging the ground to escape the wind's full force. Deciduous trees like oaks and maples drop their leaves so that they escape the winter winds, while firs and spruces keep their leaves, or needles, which are so small that they do not lose water easily.

In cold countries, the wind may have a different effect. Cold still air can feel comfortable but even a light wind can make you feel very cold. This is called a wind-chill effect and is worst where bitterly cold winds can cause frostbite on unprotected skin. Animals avoid cold winds by sheltering in burrows or in hollows close to the land. Some creatures grow chill-proof coats to insulate them. The yak has a dense coat of matted wool and a penguin has thick feathers.

This tree has been shaped by the strong winds blowing in the Windward Islands in the Carribean.

Activity: Find the force of the wind

What you will need

Two lengths of balsa wood, 12 in long and 3/8 in in cross section. Four paper or plastic cups, a knitting needle, a plastic bottle such as that used for dishwashing liquid, strong glue and clear and colored tape.

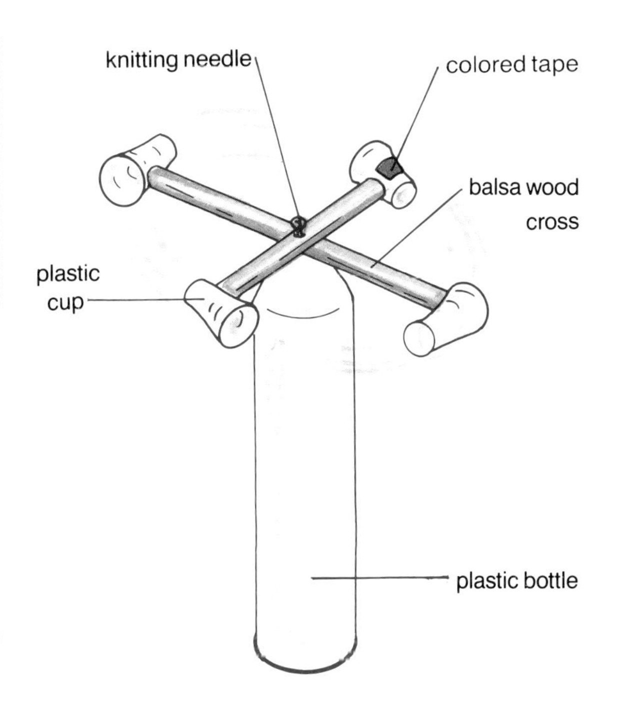

Glue the lengths of balsa wood together in the shape of a cross and carefully push a knitting needle through both at the joint. You may need an adult to help you with this. Use glue or clear tape to fasten the cups securely to the ends of the wood as shown.

Now pass the needle through the top of the plastic bottle. Make sure that the balsa-wood cross can turn freely. Finally, mark one cup with the colored tape. You have made an anemometer, an instrument that shows the force of the wind.

Take your anemometer into the wind and count the number of times it turns in one minute. Shade the anemometer from the wind. How much of a difference does this make?

Copy the table below. This is the official table of wind force, called the Beaufort scale. Use it to guess the force of the wind each day and record how many turns are equal to the force. Of course, a wind force of 8 and over will be too strong to measure with your anemometer!

Wind force	Effect of the wind on everyday objects	Number of turns
0	Smoke rises straight up.	
1	Smoke slowly drifts.	
2	Leaves start to move.	
3	Leaves move continuously.	
4	Small branches move and paper blows.	
5	Small trees in leaf sway.	
6	Branches on trees move.	
7	Whole trees sway.	
8	Twigs and small branches break off. Gale warning.	
9	Large branches break off and slight damage to houses.	
10	Trees uprooted and major damage to property.	
11	Storm conditions, very dangerous.	
12	Hurricane conditions, usually at sea or on the coast.	

9. Air and plants

Many plants use the winds so that they may reproduce. For example, pollen grains are blown from one flower to another. Even a light breeze can carry millions of pollen grains from grasses and trees. Some people are allergic to pollen and suffer from hay fever. This is caused by pollen that irritates the eyes and nose causing watery eyes and sneezing.

If a pollen grain lands on a female part of a flower then a seed will result, from which a new plant may grow. Many plants use the wind to spread their seeds. Some seeds that are blown away may have special devices to catch the wind. Dandelion and thistle seeds have small hairy parachutes. The seeds of sycamores and pines have wings that make them spin like a helicopter, slowing their fall from the parent plant so the wind can take them farther away. It is very important for these seeds to be scattered and carried for long distances in this way. This will reduce the competition for light and for water and foodstuffs in the soil when the young plants start to develop.

Some plants do not produce seeds but make spores. These are rather like seeds but they are very tiny and are released into the air by plants like mosses, ferns and fungi. They are often carried right to the top of the troposphere by rising air currents and, if caught by fast-moving winds, may be carried for hundreds of miles before the rain washes them back to the ground.

Lichens produce vast numbers of spores. When ripe, an ordinary field mushroom can release 100 million spores in an hour. A giant puffball produces even greater numbers. In its life, some scientists believe that it can produce 7,000 billion spores, shooting them into the air like puffs of brown smoke every time the wind blows against it.

A puffball releasing a cloud of tiny spores.

Activity: Catching spores

What you will need

A fresh slice of bread, a saucer, some boiling water and the use of an oven.

First make sure that there are no spores in the saucer by carefully pouring boiling water over it. Put a slice of bread into a hot oven for five minutes, also to kill any spores before the experiment begins.

Place the bread on the saucer and moisten it with a little boiling water. Then leave it on a window sill where the air from outside can reach it. Within a week you should find some fungus growing on the bread.

Examine the fungus with a magnifying glass but be very careful not to breathe in any. Are there different types of fungi on the bread?

A species of fungi called penicillium growing on an orange skin.

What this shows

Fungi use the wind to scatter their spores. Vast numbers of them travel by air and land on foodstuffs to obtain their food for growth.

A close-up picture of mold growing on stale bread. The black dots are the spores.

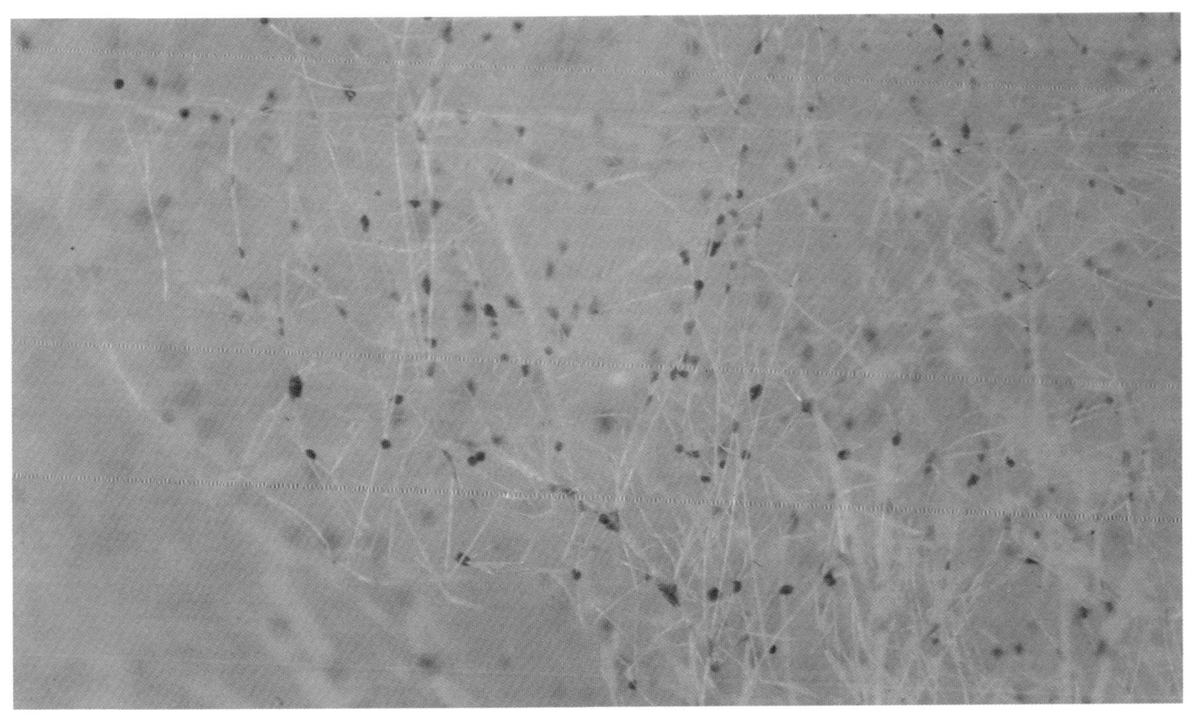

10. Animals and breathing

Animals take air into their bodies so that the oxygen can release energy from the food they have eaten. This is rather like air being necessary for fuel to burn. There are a number of different ways in which animals respire (breathe). The simplest tiny creatures take in oxygen through the surface of their bodies. Certain amphibians, like frogs, may also do this. Most fish have special organs called gills that remove the oxygen that is dissolved in water. Insects have holes along the surfaces of their bodies that open into tubes down which air passes.

Another method, that is used by mammals, birds and reptiles, is to use lungs to take in air. Lungs are special organs in which oxygen dissolves into the blood so it can be taken around the body to release energy. The result is carbon dioxide, which returns in the blood to the lungs to pass out into the air.

These animals take air into their lungs by using special muscles in their bodies. Humans use the powerful sheet of muscle called the diaphragm. This runs across the bottom of the chest making the region air tight. When the diaphragm moves down the pressure in the lungs is reduced and air is drawn into them. To breathe out the diaphragm relaxes, forcing out the air.

In addition to the diaphragm, the muscles that control the ribs are important for respiration. The ribs are the bones that surround the chest and they move upward and outward slightly as the diaphragm descends, helping to draw in air.

Animals need air to live, yet often they take in pollution in the form of dust and smoke. These go right into the sensitive lungs and harmful substances can pass into the blood. Some humans, for instance, breathe in the fumes from cigarettes, which leads to diseases of the lungs and chest.

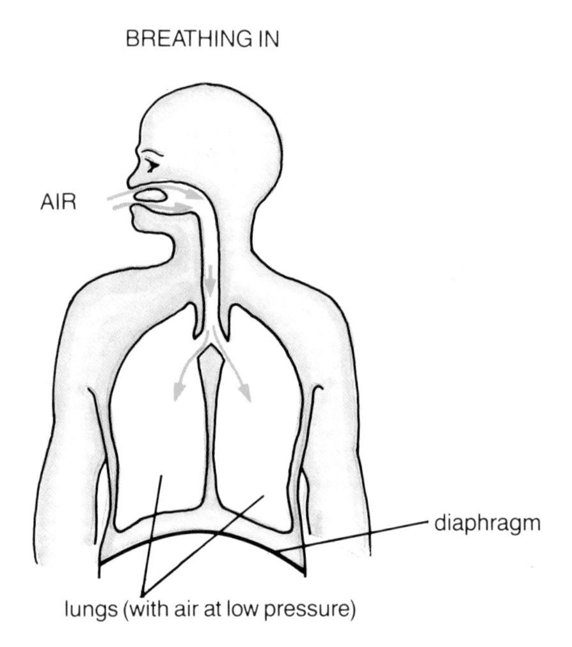

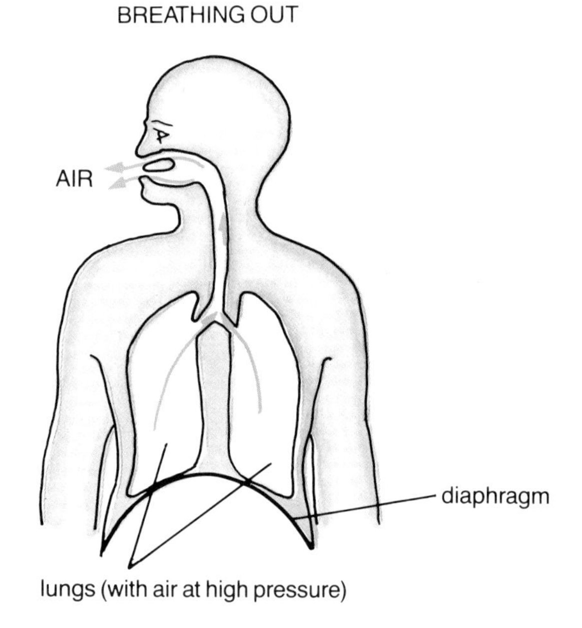

The movement of the diaphragm helps to draw air into the lungs and to push it out again.

Activity: How air is taken in

What you will need

A bell jar with a rubber stopper that has a glass tube running through it, a large sheet of rubber, small and large rubber bands, a balloon and some candle wax.

Blow up the balloon once so that it is fairly soft. Fasten it tightly to the glass tube with a small rubber band and put it into the bell jar as shown. Seal the stopper where it touches the glass with some candle wax to keep air from entering. Pull the rubber sheet over the bottom of the bell jar and firmly fix it in place with the large rubber band. Again use the candle wax to seal where the rubber touches the glass.

With the rubber sheet firmly in place, pull it in the center as far down as possible. What happens to the balloon? Why does this occur? Let the sheet go and watch what happens.

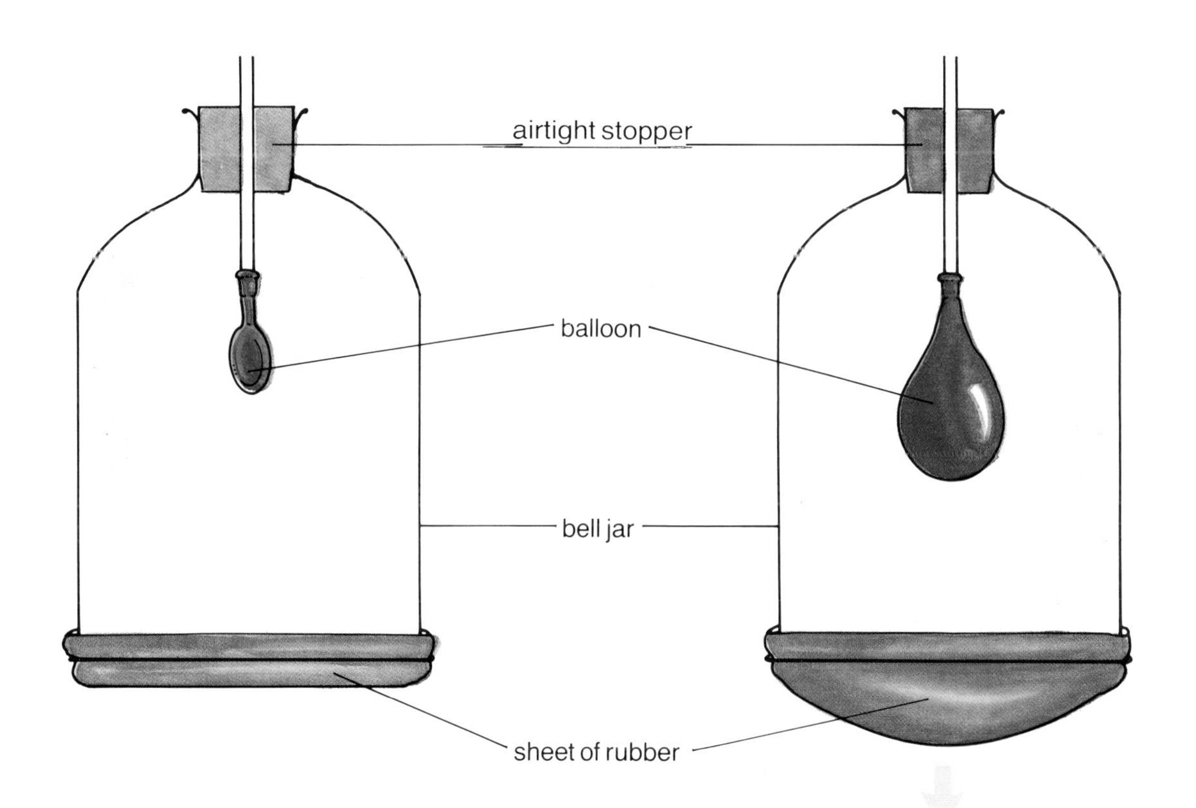

What this shows

The balloon behaves like a lung and the rubber sheet is like the diaphragm. When the rubber sheet is pulled down there is a greater amount of space in the bell jar. Air rushes into the balloon to fill the extra space. When the diaphragm is released, the air is forced out of the balloon in the same way that air is forced out of the lungs.

11. Life in the air

Flying insects are the largest group of creatures that travel through the air and they may spend most of their adult lives "on the wing." Yet they must land to lay their eggs and while the young insect is developing it cannot fly. Some insects, like certain butterflies, flutter from plant to plant and will feed, mate, lay their eggs and die in the same small area where they were hatched. Other species will spend their lives traveling, or migrating, in search of food or to mate.

The monarch butterfly is the most famous species that migrates. A large number of them live in the woodlands around the Great Lakes of North America. In the autumn, millions fly south toward Central America, a journey of 3,200 km (2,000 mi). They follow a

Dragonflies are the most agile of insects. They have four wings and hold their legs under their heads to catch the small insects on which they feed.

definite route, steering by the Sun, and rarely stopping to feed. Those arriving in Mexico, for example, assemble in specific valleys and roost in trees that have been used for generations. In the spring they mate and only then do they move north again. Few will survive the return journey, but they lay their eggs along the way.

Locusts are another type of insect that migrates, but this they do for food. North African locust swarms migrate up to 4,800 km (3,000 mi) feeding on vegetation and crops planted by humans as they go. With the help of the wind they can fly for twelve hours at a time at speeds of up to 16 kph (10 mph).

Flight is made possible by the beating of wings. A single downward stroke of a wing is enough to increase the pressure in the air below it and reduce the air pressure above it. Air moves from high- to low-pressure areas and so the creature is "sucked" upward. This is what is called lift. By repeating the downward strokes an insect can fly, and many have become experts of flight. Honeybees can beat their wings 15,000 times a minute and dragonflies can reach speeds of 30 kph (19 mph).

The bat is the only mammal that actually powers its own flight.

Insects travel at varying heights. So as not to be blown off course on windy days, butterflies fly close to the ground, sheltering behind the vegetation. However, on days with still winds they may rise to heights of 1400 m (870 ft). Apart from those that fly through the air, many insects drift. Tiny spiders spin long threads of silk to act as parachutes that catch the wind and carry them to new areas. Air currents may often take these spiders 1500 m (930 ft) but they can rise as far as twice this height. As they go, these creatures are preyed on by insect-eating birds such as swifts.

There are several mammals that often travel in the air, such as the flying squirrel, but these really glide. Only one group, the bats, actually power their own flight, fluttering through the dark to feed on insects. Some bats migrate, such as the desert bat that flies from near the Canadian border to Mexico.

12. Birds and flight

The true masters of the air are the birds. They also make long migratory flights using the winds. The arctic tern flies from the Arctic to the Antarctic every year, a round trip of 35,000 km (22,000 mi). Some birds rarely land. The wandering albatross has wings 3.5 m (11 ft) across that are difficult to flap on the ground. When at sea, however, it can swoop and soar for hours on end without a single beat of its wings.

Another group of birds that rely on gliding are the vultures of Africa. These birds spiral upward in warm rising air currents called thermals. On reaching the top of a thermal they glide down to join another farther away. In this way they can travel 100 km (60 mi) a day searching for food.

In addition to providing support while flying, most birds "row" themselves through the air with their wings. Although the work done by these wings is much more complicated than the wings of an airplane, the same basic method is used to give the lift.

The feathers that make up a bird's wing help it to be streamlined, the best shape for flight, and also provide it with airfoils. Any shape that produces lift or controls the direction of flight is called an airfoil.

The shape of a bird's wing tells you how it performs in the air. Albatrosses have long thin wings like gliders to catch the up-currents of air, and vultures have broad, squarer wings like slow-flying aircraft. Birds that need to change direction quickly need smaller wings. Peregrines, when diving at 130 kph (80 mph) sweep back their wings like high-speed aircraft. Hummingbirds are able to hover like a helicopter and can even fly backward.

A hummingbird is able to hover like a helicopter while using its long beak to take the nectar from a flower. This Anna's Hummingbird is found in California.

Activity: Make an airfoil

What you will need

A sheet of paper, tape, a drinking straw, strong thread and pencils.

Fold the sheet of paper in half and crease the fold. Slide the top half back and tape it in position so that you have a wing. Make two small holes through the wing and push the straw through so that it is held in place as shown. This also helps to hold the paper in the shape of the airfoil.

Pass the thread through the straw and tie the ends to the two pencils. Hold the pencils so that the thread is straight and move the airfoil through the air. What do you see?

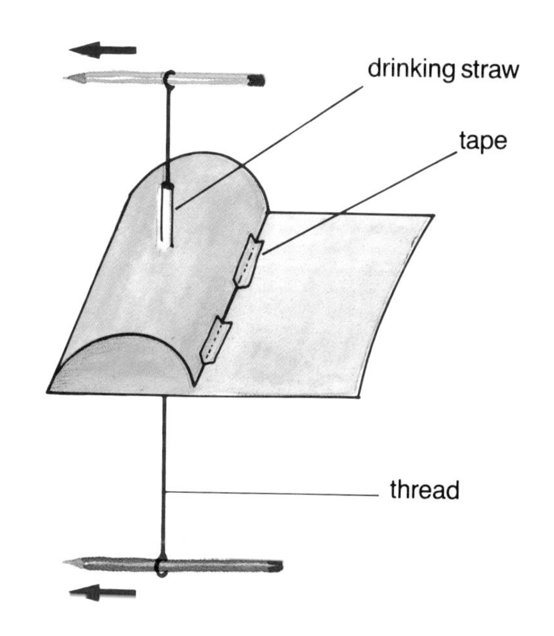

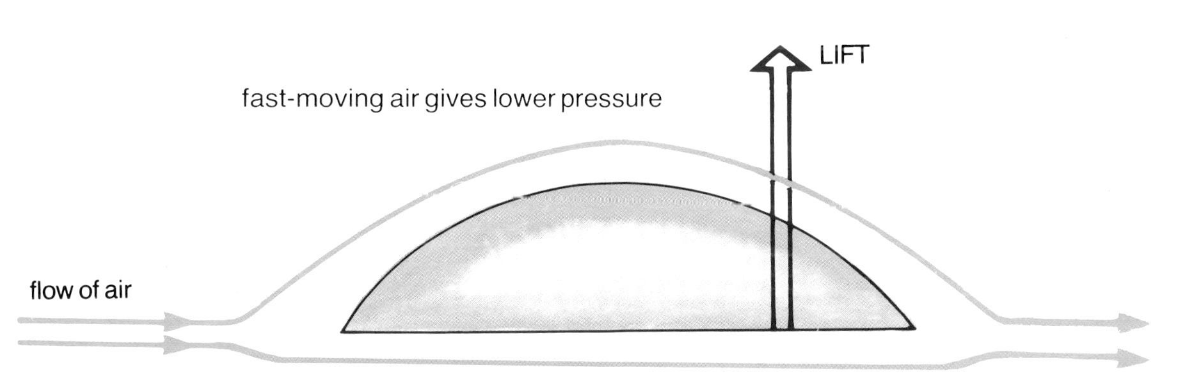

What this shows

When the air meets the upper edge of the airfoil it "splits" and flows over the upper and lower surfaces before meeting at the other end. The upper surface is longer than the lower one and so the air has farther to go. If the air splits and meets again it must travel faster over the top surface than the lower. The fast-moving air makes a low-pressure area over the top surface and a higher-pressure area over the lower one. Higher-pressure air moves to a low-pressure area so the airfoil is lifted up. This is what is called "lift" and is how a wing is pushed upward.

13. Natural pollution

It is often thought that humans are the only source of pollution. Yet there are many natural ways in which the substances scientists think of as pollutants enter the atmosphere. Dust is one example. The heavier dust from the land can be carried into the air during a storm. For example, red sand from the Sahara desert can be carried across the Mediterranean Sea and washed out of the air causing the "blood rains" in Italy.

The dust blown into the air by an erupting volcano is different. This dust is blasted into the stratosphere and thus is above the weather and cannot be washed out. Such dust may take from a few days to several weeks to be carried around the world. The dust does float down slowly and will eventually fall back to the surface, but this can take several years.

Dust in the atmosphere scatters and reflects the Sun's energy, reducing the heat and light reaching the surface. There were two large volcanic eruptions in the northern hemisphere in 1783. That part of the world cooled by 1.3 °C and it took four years for the temperature to get back to normal.

Volcanoes are also a source of major pollutants such as the oxides of sulphur and nitrogen. These gases, together with carbon dioxide, are also produced by forest fires but the greatest natural source of them is the decay of dead plants and animals. As they rot away due to the action of bacteria, carbon dioxide, sulphur products and ammonia (a gas containing nitrogen) are released.

Since these gases are spread out over the world they have little effect on life. However, the release of these gases due to human activity is concentrated in small areas and does have a definite effect.

The dust from the land may be carried into the air by a storm and washed out again by the rain, as in this mountainous region in Tanzania.

Activity: A gas released by decay

A giant plume of smoke, gas and dust is released into the atmosphere by an erupting volcano. This is Krakatoa in Indonesia, which erupted in 1981.

What you will need

About forty peas, a flask, some cotton and some red litmus paper.

Boil a little water and let it cool. Put the peas in the flask and add just enough of the water to cover them. Let the peas soak for a full day and then pour away most of the water. Make a stopper out of cotton and push it into the neck of the flask. Leave the flask in a warm place (near the radiator or in the sunlight) for about two weeks and watch what happens.

After this time open the flask and place a strip of damp red litmus paper in the air above the peas as shown. What happens? Do you notice a particular smell coming from the flask?

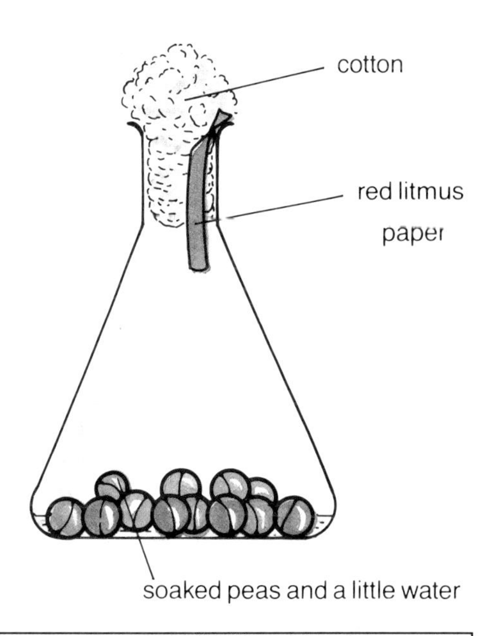

What this shows

You may have recognized the smell of ammonia. Red litmus paper turns blue in this gas. This shows that vegetation that decays naturally gives off ammonia.

14. The human polluters

Although vast amounts of materials and gases enter the atmosphere naturally the most polluting ones come from human activity. In the last 150 years much pollution has been produced by the development of industries. For example, the burning of coal and oil and the work of factories produces great quantities of dust and soot that settle on buildings or are washed out by the rain.

Sulphur-containing fuels produce sulphur dioxide gas when burned in power stations and factories. The yellow London smogs were a result of this gas combining with fog and smoke. Such a smog resulted in the death of 4,000 people in five days in 1952. The exhaust from cars has also caused a bad type of smog in more recent years, cutting down the light from the Sun by up to 90 percent in many cities.

Another substance entering the air in car exhausts is lead, which is used in gasoline to make cars run better. In Philadelphia, near-poisonous levels of this metal were found in the blood of traffic policemen and birds. As a result of such findings regulations have been passed to reduce the amount of these substances introduced into the air through burning fuels.

An alternative to these fuels is to use nuclear power to produce electricity. Yet there is much concern over the safety of this kind of power supply. In 1986 there was a severe accident at a nuclear power station in Chernobyl in the U.S.S.R. The resulting cloud of radioactive particles spread across Europe, carried by the wind.

There was great concern over the radioactivity being washed out of the air by the rain and entering the grass, there to be eaten by cattle, which would pass the substances to humans in milk and meat. The effects of the accident will only become clear in the years to come.

The air pollution from this industrial plant in Britain has killed some of the surrounding trees.

Activity: Make a pollution collector

What you will need

Several funnels, beakers or jars, filter papers and a magnifying glass or microscope.

Make several pollution collectors by placing a funnel with filter paper in each beaker. Leave the collectors in various places on a rainy day and find the direction of the wind by watching the clouds.

When the rain has passed remove the filter papers, open them out and let them dry. Examine the papers with a magnifying glass or a microscope. Are there any particles of pollution like soot? Where do they come from, remembering the wind direction? Are there any natural substances like pollen or spores?

Repeat the experiment on a dry day. Do you collect more or less particles when it does not rain?

What this shows

The air is full of particles, many of which are pollutants, and the rain washes out a great many of them.

Pollution caused by a natural gas plant in New Zealand.

15. Rain turns to acid

When fossil fuels are burned sulphur and nitrogen oxides are released. These gases are always in the air from natural sources, but most are produced by power stations, industries and motor vehicle exhausts.

If these gases are swept up into the air by tall chimneys they can travel great distances before returning to the surface. The longer they stay in the atmosphere, the more likely they are to combine with the moisture in the air to form sulphuric and nitric acids. It is these that fall to the ground as acid rain.

Because of the air system the gases may be carried to neighboring countries, and thus acid rain is often exported from one country to another. The effects of acid rain were first noticed in Sweden, with some 10 percent of that country's lakes no longer containing any live fish. This increase in the acidity of lakes is found throughout Europe and particularly in the eastern U.S. and Canada.

As well as affecting life in lakes, acid rain enters the soil and stunts the growth of plants. Lichens are very sensitive to sulphur dioxide in the air and can be used as indicators of the gas. Lichen "deserts" have been found around large towns where very sensitive types of lichens do not grow.

It is not known to what extent acid rain is to blame for the damage of about half of West Germany's forests. However, it is known that acid rain can remove nutrients from the soil and damage roots and leaves. Some scientists believe that other pollutants such as certain smogs containing the gas ozone may be responsible.

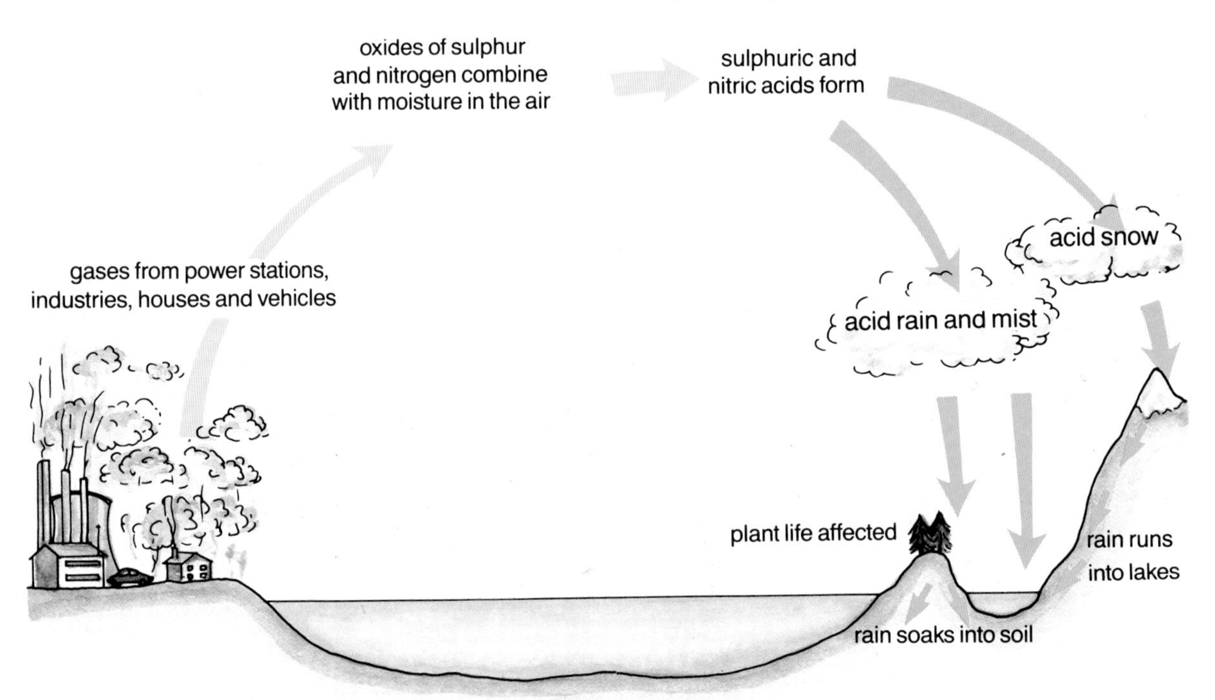

When we burn fossil fuels in factories, homes and vehicles, gases are given off. Among these gases are the oxides of sulphur and nitrogen, which combine with the moisture in the air to provide acid rain or snow.

Activity: Lichens and pollution

What you will need

A magnifying glass, a map of the local area, paper and pencils.

Lichens are types of plants that grow on trees and stone. Some are very sensitive to sulphur dioxide gas in the air. For example, the shrub types, which look like miniature gray-green bushes without leaves, are found only in remote areas.

The leafy kind are also gray-green in color and may be found in areas with a little pollution.

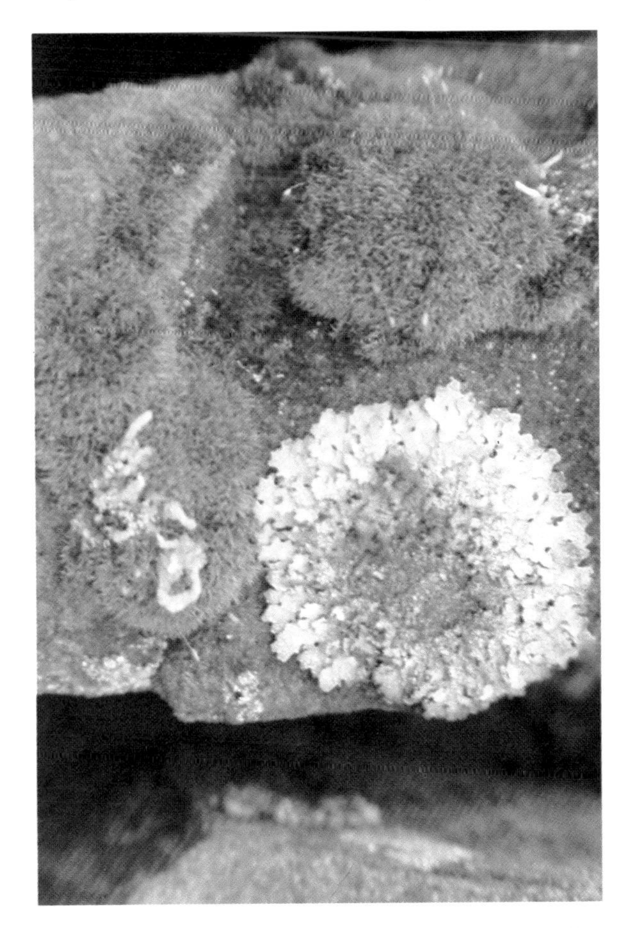

The leafy type of lichen, which is tolerant of a little air pollution.

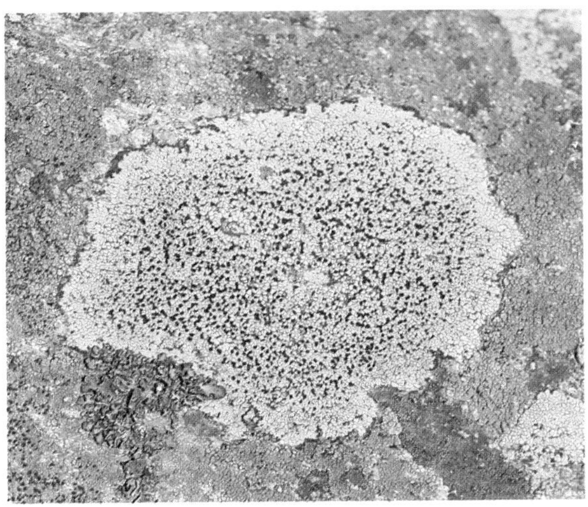

The flat type of lichen, which is very tolerant of pollution.

They have flat leaves and grow out slightly from the surface of the stone. The flat type are orange and green in color and grow flat against the surface. They are very tolerant of sulphur dioxide, but no lichens will be found in places that have very great amounts of pollution. There will be a lichen desert.

Look at the buildings in your area, especially old ones, for lichens on walls, roofs and gravestones. Copy a map of the local area and mark where you find lichens. Show which type they are and how many of them you find, noting if they are rare, if there are some, or many.

If possible visit the center of the town. Are any lichens found there or is it a lichen desert? Why do you think lichen deserts occur? Move away from the town center and continue to look carefully. Do lichens become more frequent?

What this shows

The results of your investigation show the extent of sulphur dioxide pollution in the area.

16. Harnessing the wind

Humans need energy for heat, light and everyday life and we get most of this from fossil fuels like coal and oil as well as nuclear fuel. Yet using fuels creates problems such as acid rain, smog and radioactivity. There is also a limit to the amount of fuel available. For example, the sources of oil and gas are running out.

In the face of these problems scientists are seriously looking at energy sources that do not use up fuels. These are called "renewable resources" and the wind is one example.

Using wind power is not a new idea. The Egyptians used sails on their boats 5,000 years ago and windmills were used by the Persians to grind corn 2,500 years ago. More recently windmills have been used to run generators that produce electricity.

Different designs of wind generators are being tested, including giant ones with sails 80 m (50 ft) across as well as "wind farms" made up of many smaller versions. In the U.S. wind power is now used in many areas to provide electricity for thousands of homes, and in the U.S.S.R. there are plans to have 150,000 wind generators in use by 1990.

However, some people say that large wind generators would be an eyesore and that they would be very noisy. This would be especially so since a great many would be needed to do the work of a coal or nuclear power station. For the most part, though, the effect of wind generators on the environment would be small compared to these sources of energy and there would be no need to buy or dig for fuel. It seems very likely that wind power will play an increasingly important part in providing some of our energy needs.

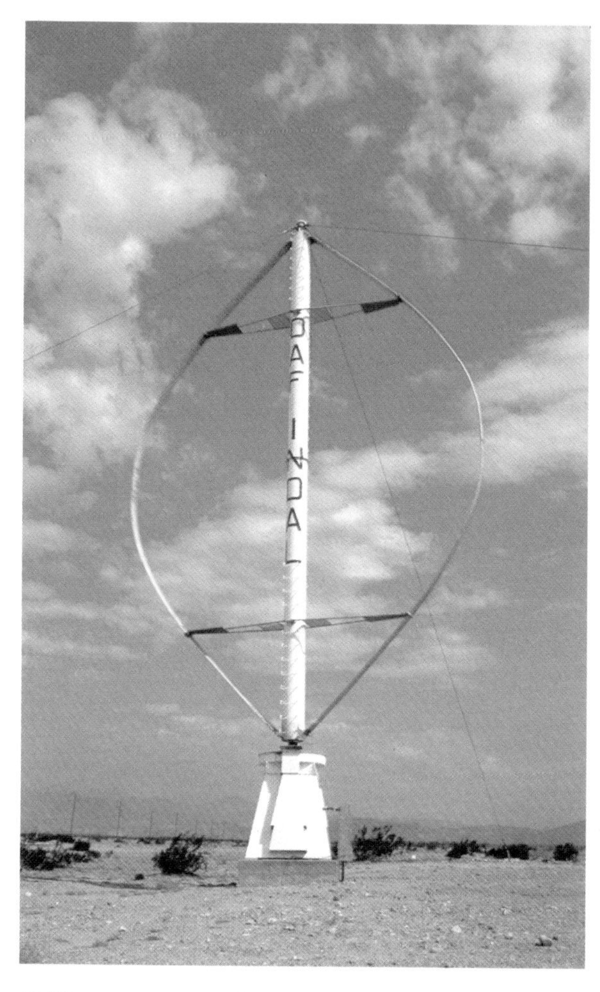

Different types of wind generators, such as this Darrieus one, are being tried out.

Windmills are not a modern invention. This one was built in the mid-nineteenth century.

Activity: Make a windmill

What you will need

A cork, a piece of balsa wood, a short piece of thick wire, two glass beads, thread and cardboard.

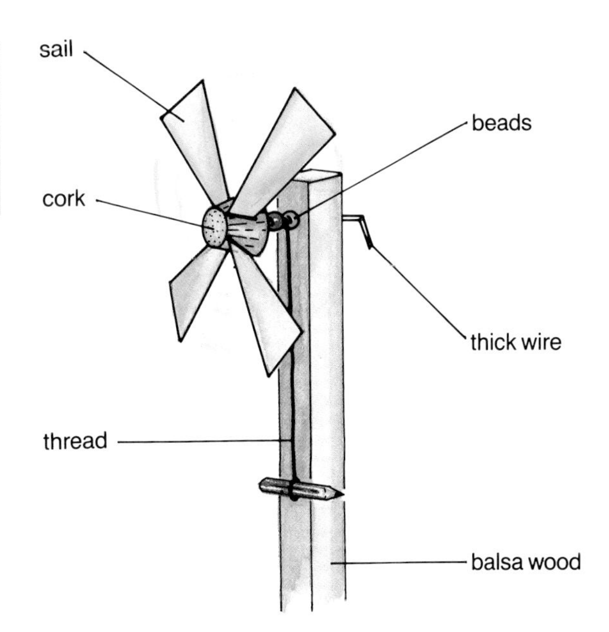

Carefully cut four slits in the cork at a slight angle for the sails, as shown. Push one end of the wire into the cork and thread the beads onto it. Tightly tie the end of a length of thread between the two beads. Then make a hole through the balsa wood and pass the wire through it. Make sure that it can turn freely. Bend the free end of the wire so that it will not fall out.

Cut four sails of the same shape from the cardboard so one end will fit into a slit in the cork. The windmill is now ready to do some work. Tie a lightweight object like a small pencil to the thread and blow on the sails or take it into a breeze.

Remove the cork and replace it the other way around. Which way does the windmill move now? Make sets of sails of different shapes such as long narrow ones or wide fanlike types. How do these perform?

What this shows

The wind can be used to do some work.

Propeller wind turbines on a "wind farm."

17. Upsetting the balance

The weather changes each day, but the climate normally remains the same for a long time. Yet great changes in the climate are possible. About 225 million years ago the Earth was a much hotter place and this favored the development of the great reptiles – the dinosaurs. These were the most successful creatures on the Earth for 160 million years until there was a change in the environment and they slowly died out. This change may very well have been the cooling of the climate.

There can be thousands of years between changes of climate. The last ice age ended 10,000 years ago and was followed by periods of warming and cooling until today, when some scientists think we are due for another ice age. Such a natural change in the climate, however, may not be as important as a man-made one. Humans are affecting the warmth, winds and rainfall of the Earth.

Much of the energy from the Sun is reflected back into space by the surface of the planet. Areas with little vegetation reflect most of this energy while areas like the rain forests reflect little. In recent years great areas of these forests have been cut down or burned, providing wood for building and space for new settlements. Many scientists believe that the clearing of the rain forests will cause an increase in the "shininess" of the Earth with more heat reflected than before. This could result in changes to the movement of the air above the forests, which might upset the air system around the world. North America, Europe and the U.S.S.R. might then receive less rainfall, and India and parts of the Sahara desert in Africa might receive more.

There is another way in which the natural balance of the Earth's climate may be upset. Carbon dioxide keeps the planet warm by the so-called "greenhouse effect." In the last 150 years, however, the amount of carbon dioxide in the atmosphere has increased due to human activity. Burning fossil fuels in power stations and factories and the exhausts of motor vehicles release huge quantities of this gas into the air, as does the burning of the rain forests.

It is possible that a warming of the atmosphere will occur as a result. This could lead to great changes in the climate. Rainfall patterns might change, with many countries

The burning of the rain forests add to the build-up of carbon dioxide in the atmosphere. Such a build-up may increase the "greenhouse effect" and lead to a warming of the atmosphere.

Ice floating on the Bellinghausen Sea off the Antarctic coast. A warming of the atmosphere due to an increase in the greenhouse effect might cause the poles to start melting.

receiving more rain and others receiving less. It is possible that a build-up of carbon dioxide would also encourage photosynthesis in plants, balancing the effect of the lack of rain in some countries. This could lead to an increase in the production of crops, which in turn would absorb the extra carbon dioxide.

Yet, there is another possible result of a carbon dioxide build-up. The warming of the atmosphere might lead to the melting of the poles with the sea level rising as a result. Over many years the sea would rise to flood coastal cities and low-lying areas. This would mean that these cities would have to be rebuilt inland at great cost.

Other gases that add to the "greenhouse effect" are those used in some spray cans and fire extinguishers. Life on Earth is protected from the Sun's ultraviolet radiation by the ozone layer, and it seems possible that those gases released by such sprays could lead to the ozone being damaged. If more ultraviolet radiation hit the Earth's surface much life would be harmed and plant growth slowed. It seems that anything that may upset the natural balance of the air must be treated with great care.

18. Talking points

There have been great successes in controlling pollution. Many towns suffer from smogs that would be worse if it were not for special devices fitted to vehicle exhausts. Does your town suffer from smog? Lead is another dangerous substance released by vehicle exhausts. The U.S. was the first country to make laws lowering the amount of lead in gasoline. Many other countries have begun to make gasoline "lead-free."

Some people have argued that the cost of making vehicle exhausts safer is very great.

Trees affected by acid rain in the Black Forest, Germany. The gases from power stations, factories and vehicles are thought to be responsible.

Do you think that the people who buy vehicles would be happy to pay extra for safer exhaust fumes? Perhaps people may be encouraged to use vehicles less. Would people be willing to use buses and bicycles more? What are the advantages of doing this in large cities?

When countries reduce the amount of pollution in the air above them they often send much of it to their neighbors. For example, power stations have very tall chimneys, which take the pollution away from the local area but often help to cause acid rain elsewhere. Is this a fair way to deal with the acid-rain problem?

Attempts have been made to reduce the effects of acid rain. In Sweden huge amounts of lime are dropped into affected lakes at great cost, but even more is necessary to correct the damage. Is this the best way to control the effects of acid rain? Would it make more sense to stop the polluting gases from escaping in the first place?

New air pollution laws have already reduced the amount of sulphur dioxide gas released into the air. Special equipment like that used to remove sulphur and nitrogen oxide gases from burning coal are not cheap, though. The cost would be paid for by governments, industries and the public. Do you think that everyone would be willing to pay more for electricity if all coal-burning power stations were fitted with such equipment?

About twenty European countries have joined together, including those worst affected by acid rain, promising to reduce the amount of the gases released into the air by 30 percent. Yet some countries, like Britain, have refused to join. Britain is the largest producer of sulphur dioxide in Europe and its government has said that it does not intend to pay for expensive controls on gases when there is so much uncertainty about the effects of acid-rain. Is this the most sensible way a government can react to the acid-rain issue?

Instead of burning coal, more nuclear power could be used for our energy, though this may also have dangers. There is much worry over the safety of nuclear power as well as its high cost. For instance, the United States has not built a nuclear plant since an accident at one in 1979. Do you think that it would be wise for all countries to stop building nuclear power stations and rely on other sources of energy, like coal and the wind? Or would it be best to use all sources of energy in case one type ran out or had to be given up? What bad effects would wind generators have on the environment? Maybe one of the best things to do would be to use less energy. How would you reduce the amount of energy you use and waste each day?

Industries, vehicles and cutting down the rain forests release carbon dioxide into the air. This may increase the "greenhouse effect" and lead to the atmosphere's warming up. However, much remains unknown about this effect. Is it a good idea to try to control the amount of this gas released into the air until we know more? How may it be possible to reduce the amount of this gas?

The replanting of trees has been suggested to soak up the extra carbon dioxide in the air. An enormous number of trees would be needed and this would be very expensive. Yet the cost, however large, may be small compared to the possible results of an increased greenhouse effect, such as the melting of the polar ice and a rise in sea level.

Tree seedlings in a nursery in Himachal Pradesh, India. They have been planted as part of an attempt to reforest the Himalayas where vast numbers of trees have been cut down.

Glossary

Absorption The taking in of a substance or energy.

Acid rain Rain is normally slightly acid. Yet gases like the oxides of sulphur and nitrogen, which are released into the air by burning fossil fuels, dissolve in the water in the air and fall as acid rain, snow or mist.

Airfoil An object that produces lift.

Alga(e) A type of plant that contains chlorophyll but lacks stems, roots and leaves.

Ammonia A gas that contains nitrogen and is released by the decay of plants and animals.

Argon A gas that makes up less than 1 percent of the atmosphere.

Atmosphere The layer of gases that surrounds a planet, held there by gravity.

Bacteria Extremely small living things that bring about the decay of plant and animal remains and wastes.

Carbon dioxide A colorless gas that makes up 0.03 percent of the air. It is released when fossil fuels and wood are burned. Respiration by living things also releases this gas.

Chlorophyll The green chemical in plants that absorbs the light energy required for photosynthesis.

Climate The main weather conditions in an area over a long period of time.

Condensation The change of a vapor into a liquid.

Cycle A series of events, repeated regularly, in the same order.

Deciduous Shedding leaves each year at the end of the growing season.

Ecology The study of how living things affect, and are affected by, their environment.

Energy The power to do work. There are several different types of energy. Heat and light energy are produced by the Sun.

Environment The world around us, or our surroundings, including all living things. The place where an animal or plant lives may be called its environment.

Evaporation The change of a liquid into a vapor.

Filter paper A material like blotting paper that allows liquids to pass through it but prevents solid particles from doing so.

Fossil fuels Those fuels (oil, gas and coal) that have been formed in the ground over millions of years from the decay of once living things.

Fungi Simple plants that do not contain chlorophyll.

Generator A machine that produces electricity from mechanical energy.

Greenhouse effect An effect caused by carbon dioxide and other gases whereby heat is trapped in the atmosphere and prevented from escaping back into space.

Ionosphere An outer layer of the atmosphere that contains only 2 percent of the Earth's air.

Lead A metal that is poisonous to life.

Lichen A type of plant consisting of an alga living in a fungus. Lichens are found as crusty patches or bushy growths and some are good pollution indicators.

Nitrogen The gas that makes up 78 percent of the atmosphere.

Oxygen The gas that makes up nearly 21 percent of the atmosphere. It is essential for life.

Ozone A gas that is a form of oxygen. It is poisonous and is produced in certain smogs by the action of sunlight.

Ozone layer The layer of ozone in the stratosphere that absorbs ultraviolet rays from the Sun.

Photosynthesis The food-making process carried out by green plants. The Sun's energy is absorbed by chlorophyll in plants to make food from carbon dioxide and water.

Pollen The fine, powdery substances produced by some plants. Pollen combines with the female parts of a flower to develop into seed.

Pollution The release of substances into the air, water or land that may upset the natural balance of the environment. Such substances are called pollutants.

Radioactive Giving out harmful rays.

Rain forest A dense forest found in the hot, tropical areas of the world.

Renewable resource A source of energy that does not need fuel and cannot be used up, such as the winds, moving water and the Sun.

Respiration The process in which oxygen is taken in from the air to release energy and carbon dioxide from food.

Smog A harmful mixture of fog, smoke and other gases that may form in the air above towns. Some smog contains ozone as a result of the action of sunlight.

Stratosphere The layer of the atmosphere that extends from 12 km (7 mi) above the surface to a height of 150 km (90 mi).

Sulphur dioxide A gas released when sulphur-containing fuels like coal are burned.

Troposphere The layer of the Earth's atmosphere nearest to the surface where all life is naturally found.

Ultraviolet radiation A dangerous form of energy that is deadly to most life on the planet's surface. The Earth is shielded from it by the ozone layer in the stratosphere.

Water cycle The circulation of the Earth's water, in which water evaporates from the sea into the atmosphere, where it condenses and falls as rain or snow. The water returns to the sea in rivers or evaporates again.

Water vapor Water in the form of a gas.

Wind-chill effect The chilling that occurs when a cold wind removes body heat.

Further information

Books to read

Acid Rain by Katherine Gay. Franklin Watts, 1983.
Ecosystems and Food Chains by Francene Sabin. Troll Associates, 1985.
The Future of the Environment by Mark Lambert. Bookwright, 1986.
Pollution bt Herta S. Breiter. Raintree Publisher, 1978.
Pollution by Geraldine and Harold Woods. Franklin Watts, 1983.
Pollution: The Noise we Hear, by Claire Jones *et al*. Lerner Publications, 1972.
Weather, Electricity, Environmental Investigations by Sandra Markle. The Learning Works, Inc. 1982.

Organizations to Contact

Audubon Naturalist Society of the Central Atlantic States
8940 Jones Mill Road,
Chevy Chase, Maryland 20815
301–652–9188

Children of the Green Earth
P.O. Box 200
Langley, Washington 98260
206–321–5291

Clean Water Action Project
317 Pennsylvania Avenue
Washington, D.C. 20003
202–547–1196

The Conservation Foundation
1717 Massachusetts Avenue, N.W.
Washington, D.C. 20036
202–797–4300

Environmental Action Foundation
1525 New Hampshire Avenue, N.W.
Washington, D.C. 20036
202–745–4870

Environmental Defense Fund
257 Park Avenue South, Suite 16
New York, New York 10016
212–686–4191

Greenpeace, USA
1611 Connecticut Avenue, N.W.
Washington, D.C. 20009
202–462–1177

National Audubon Society
950 Third Avenue
New York, New York 10022
212–546–9100

National Wildlife Federation
1412 16th Street, N.W.
Washington, D.C. 20036
202–797–6800

World Watch Institute
1776 Massachusetts Avenue, N.W.
Washington, D.C. 20036
202–452–1999

World Wildlife Fund
1255 23rd Street, N.W.
Washington, D.C. 20037
202–293–4800

Index

Picture acknowledgments

The author and publishers would like to thank the following for allowing their illustrations to be reproduced in this book: Ardea 11 (*bottom*); David Bowden *frontispiece*, 16; Bruce Coleman Limited *cover* (*bottom* E. Crichton, *left* K. Tayloir, *right* N.G. Blake), 6 (NASA), 16 (F. Lanting), 18 (R.P. Carr), 20 (J. Fry), 26 (F. Sauer), 27 (K. Taylor), 28 (B. & C. Calhoun), 30 (R.I.M. Campbell), 31 (D. & M. Plage), 33 (C. Molyneux), 40 (H. Reinhard); Earthscan 38; Cecilia Fitzsimons 7, 8, 9, 13, 19, 21, 24, 25, 29, 31, 37; G.S.F. Picture Library 22, 23 *(both)*, 35 *(right)*, 39; Jimmy Holmes 35 *(left)*, 41; Southern California Edison Company 36 *(right)*, 37 *(bottom)*. All other illustrations from Wayland Picture Library.